Like A Lizard

April Pulley Sayre

Illustrated by

Stephanie Laberis

BOYDS MILLS PRESS
AN IMPRINT OF HIGHLIGHTS
Honesdale, Pennsylvania

Can you

American badger

run like a lizard?

six-lined racerunner

Sun like a lizard?
Gila monster

Bob your head like a lizard?

One, two!

desert spiny lizard

flying dragon

Namib dune gecko

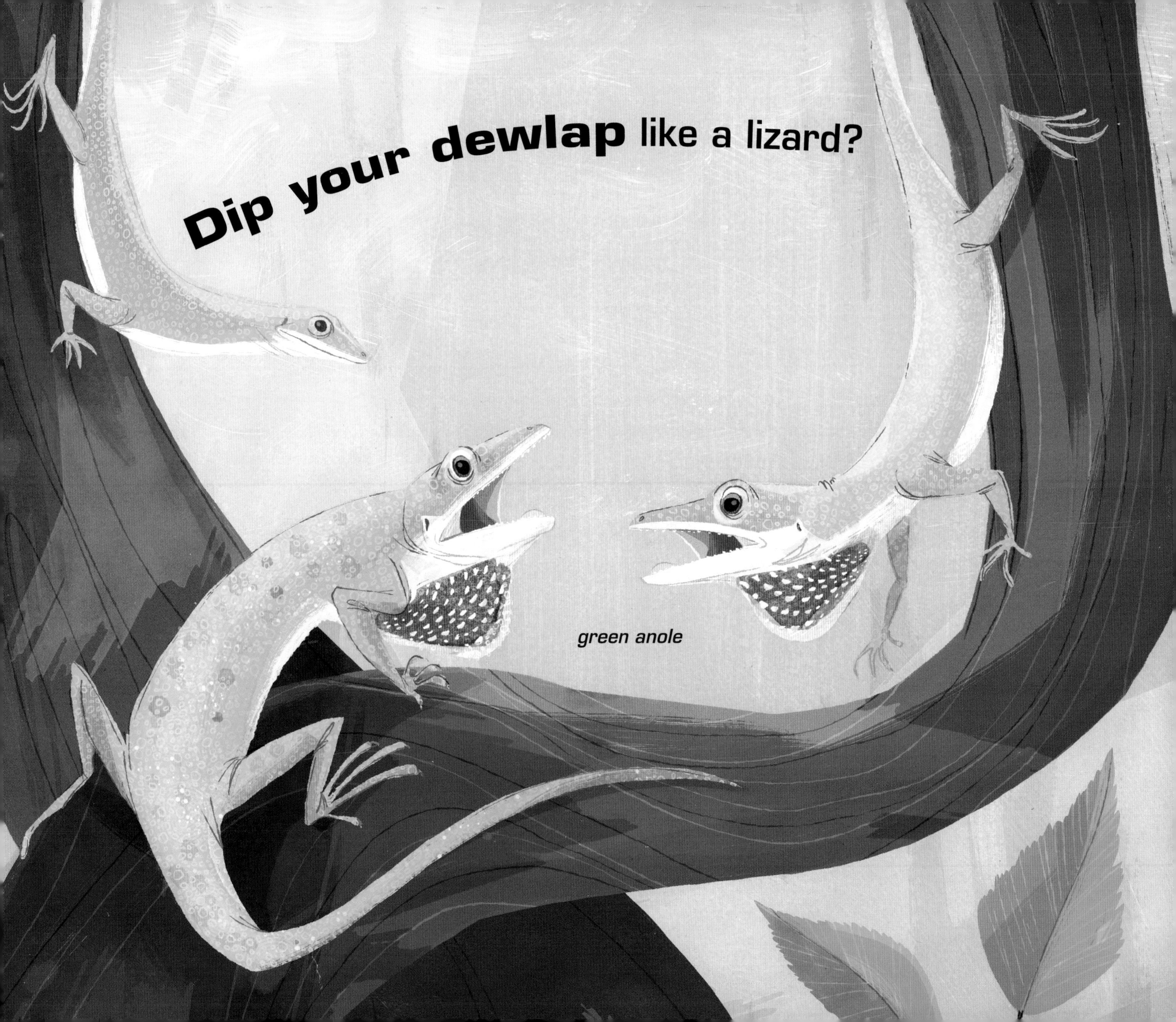
Dip your dewlap like a lizard?
green anole

Show blue?

shingleback lizard

Could you **drape** like a lizard?

green iguana

Gape like a lizard?

frill-necked lizard

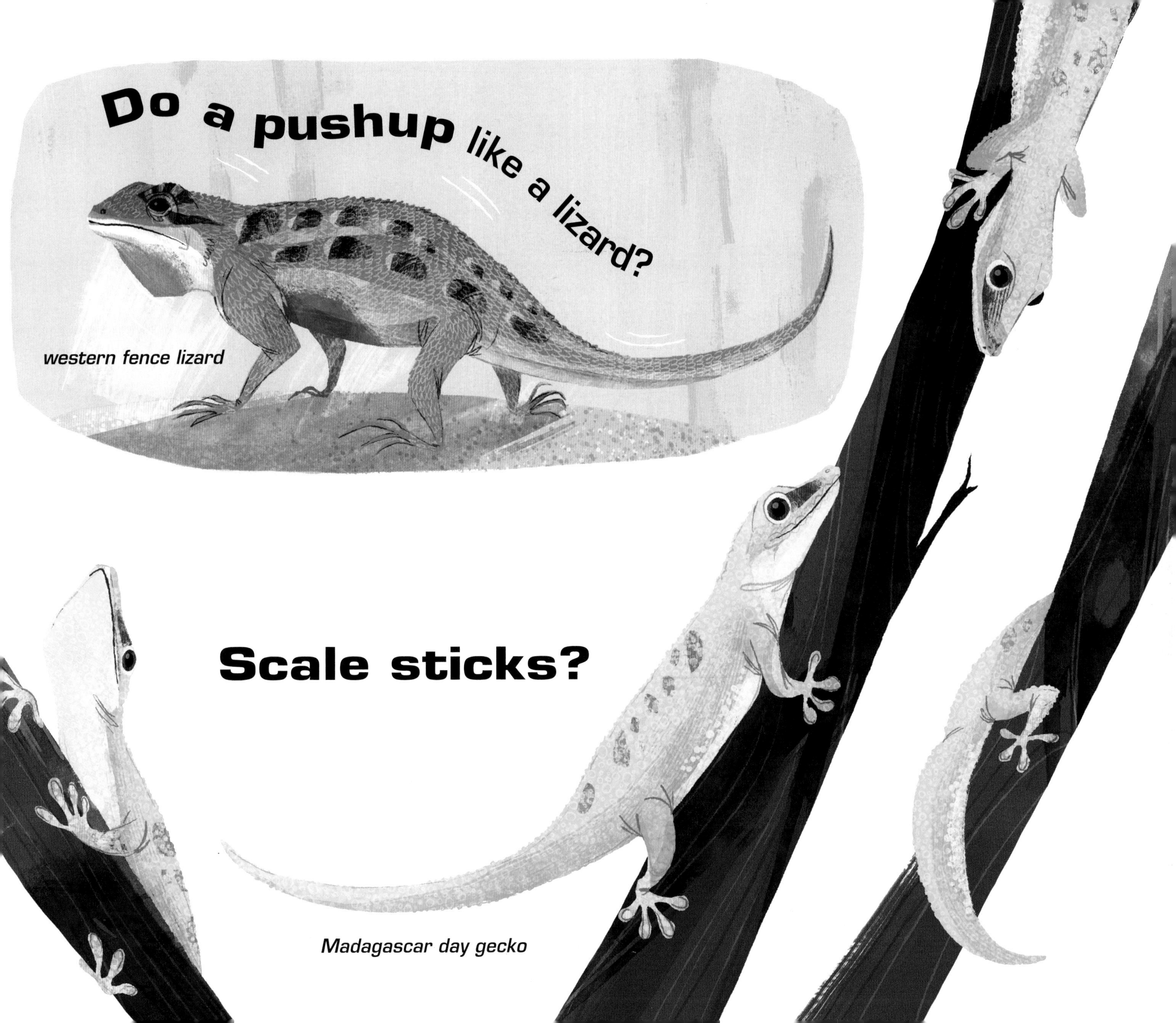
Do a pushup like a lizard?
western fence lizard
Scale sticks?
Madagascar day gecko

Curl in like a lizard?

western osprey

armadillo girdled lizard

Shed skin like a lizard?
leopard gecko

Smell well like a lizard—

Komodo dragon

Flick licks?

Would you

lunge like a lizard?
Asian monitor lizard
buff-striped keelback

Sponge like a lizard?
tokay gecko

Chew bugs like a lizard?
red-headed rock agama
Mouth mash?
land iguana

Blink like a lizard?
slender glass lizard

Drink like a lizard?

Texas horned lizard

Cross water like a lizard—
eyelash viper

Tail dash?
green basilisk lizard

Will you
hide like a lizard?
chuckwalla

Swim side-to-side like a lizard?
marine iguana

Cool your feet like a lizard,

Two-by-two?

shovel-snouted lizard

Just don't

hatch like a lizard!

veiled chameleon

No need to match like a lizard.

satanic leaf-tailed gecko

Pretend—

thorny dragon

coyote

don't **defend** like a lizard . . .
greater short-horned lizard

Be you!

More About Lizards

Could you live like a lizard? If you did, you'd have many choices of how to behave. About 6300 species, or kinds, of lizards live on Earth. Some are slow walkers. Others are fast runners. Some live in burrows. Others live in trees. Some lay eggs. Others give birth to live young. Lizards live in rain forests, grasslands, deserts, and on the shores of seas and lakes. They're more common in warmer habitats.

Like turtles, crocodiles, and snakes, lizards are reptiles. They have backbones and scales. They're cold-blooded, which means that their body temperature varies with the temperature of the environment. Lizards often warm up in the sun before they become active. They cool down in shade or in a burrow.

Many kinds of lizards are abundant; others are rare. Habitat loss is the biggest threat to most lizard species, though the pet trade is a danger, too, since wild lizards may die in transit when they're captured and shipped. (Some pet lizards are bred in captivity, not taken from the wild.)

Lizards and people, however, can live in harmony. Many people, especially in the tropics, welcome wild geckos into their homes. The geckos provide free pest control by gobbling up cockroaches, mosquitoes, and other insects.

More About the Lizards in This Book

Unlike mammals, reptiles keep growing and growing. Their growth slows down when they're adults but never stops. So we've listed the typical maximum total (including tail) lengths of each species. There are a few biggies!

Run

Six-lined racerunner (*Aspidoscelis sexlineata*)
Eastern and central United States
Length: 9.5 inches (24 centimeters)
Six-lined racerunners bask in sunshine on rocks and logs. The warmer they are, the faster they run.

Sun

Gila monster (*Heloderma suspectum*)
Southwestern United States and Mexico
Length: 21.5 inches (55 centimeters)
Gila (HEE-la) monsters are slow-moving lizards. They are one of the few venomous lizard species. Scientists are studying Gila venom—the poison they inject when they bite—to see if it might be useful for treating diseases.

Bob your head

Desert spiny lizard (*Sceloporus magister*)
Southwestern United States and Mexico
Length: 13 inches (33 centimeters)
Desert spiny lizards live among rocks and sometimes in trees. Males bob their heads to attract females and to defend their territories from other males. If a desert spiny lizard bobs its head at you, it might be a warning to stay away.

Swoop

Flying dragon (*Draco volans*)
Southern India and Southeast Asia
Length: 8 inches (20 centimeters)
Active in the daytime, the flying dragon eats mostly ants and termites. It opens the folds of skin on its sides and uses them to glide, in a long swoop, from a tree to the ground. It steers with its tail.

Scoop

Namib dune gecko (*Pachydactylus rangei*)
Namib Desert and southwestern Africa
Length: 5.5 inches (14 centimeters)
This nocturnal lizard, also known as the web-footed gecko, is one of the few animals able to live in the Namib Desert. Its large eyes help it hunt crickets and beetles in the dim light of the stars or moon. With its webbed feet, it can scoop sand away and burrow beneath it to escape the daytime heat.

Dip your dewlap

Green anole, also called the **Carolina anole** (*Anolis carolinensis*)
Southeastern United States
Length: 8 inches (20 centimeters)
Green anoles change color from green to brown, but not for camouflage. Their change is related to temperature and activity. Male anole lizards puff out their dewlaps, skin flaps on their throats, to attract females and to intimidate other males.

Show blue

Shingleback lizard (*Tiliqua rugosa*)
Australia
Length: 22 inches (56 centimeters)
This member of the skink family is also known as the sleepy lizard, stumpy-tailed lizard, or pinecone lizard. A shingleback lizard sticks out its bright blue tongue to scare predators.

Drape

Green iguana (*Iguana iguana*)
Southern Mexico, Central America, South America, and the Caribbean
Length: 6.6 feet (2 meters)
Green iguanas spend most of their time in the treetops. They eat mainly green plants and fruit. Basking in sunlight helps them digest food. They drape their bodies over branches, often hanging above water. Good swimmers, they leap into water at signs of danger.

Gape

Frill-necked lizard (*Chlamydosaurus kingii*)
Australia and New Guinea
Length: 3 feet (0.9 meters)
The frill-necked lizard is named for the colorful frilly collar of folded skin around its neck and shoulders. When alarmed, the lizard gapes, showing its bright yellow mouth, and spreads out its frill. It may also raise up and hiss in warning.

Do pushups

Western fence lizard (*Sceloporus occidentalis*)
Western United States and northwestern Mexico
Length: 8 inches (20 centimeters)
Western fence lizards often bask on fences and rocks. They bob their heads and do "pushups," probably to defend their territories from other males.

Scale sticks

Madagascar day gecko (*Phelsuma madagascariensis*)
Eastern Madagascar
Length: 10 inches (25 centimeters)
Madagascar day geckos hunt crabs, insects, spiders, and scorpions. They eat fruit and nectar and drink dew from leaves. Their wide toe pads help them climb just about any surface: sticks, palm leaves, bananas, rocks, tents, and houses!

Curl in

Armadillo girdled lizard (*Ouroborus cataphractus*)
West coast of South Africa
Length: 8 inches (20 centimeters)
This African lizard lives in dry, open areas. When threatened, it curls into a circle and grasps its tail in its mouth. The sharp, spiky scales covering its neck, body, and tail face outward as a defense.

Shed skin

Leopard gecko (*Eublepharis macularius*)
Afghanistan, Pakistan, northwestern India, and Iran
Length: 10 inches (25 centimeters)
Leopard geckos are named for the spots that develop as they mature. They live in dry, rocky areas. After leopard geckos shed their skin, they eat it.

Smell well, flick licks

Komodo dragon (*Varanus komodoensis*)
Komodo Island and several other islands in Indonesia
Length: 10 feet (3 meters)
Indonesia is home to the world's largest living lizard, the Komodo dragon. The Komodo dragon has strong legs, sharp claws, and a venomous bite. It eats prey as large as deer and water buffalo. Like other lizards, the Komodo dragon flicks out its tongue to sample the air. The tongue pulls air particles back into its mouth where the Jacobsen's organ on the roof of its mouth tastes and smells the air.

Lunge

Asian water monitor (*Varanus salvator*)
Southern and Southeast Asia
Length: 6.8 feet (2 meters)
Water monitors are large lizards closely related to Komodo dragons. They chase and lunge at prey. Good swimmers, they often hunt frogs, fish, small birds, and crocodile hatchlings. They also eat snakes, rodents, monkeys, and small deer.

Sponge

Tokay gecko (*Gekko gecko*)
Southeast Asia
Length: 14 inches (36 centimeters)
Tokay geckos do not have eyelids. Their eyes are covered by clear scales, which they lick clean with their spongy tongues.

Chew bugs

Red-headed rock agama (*Agama agama*)
Sub-Saharan Africa
Length: 10 inches (25 centimeters)
Male red-headed rock agamas develop vivid colors that brighten when they warm up in the sun. Other names for this lizard include common agama and rainbow lizard. This lizard is very unusual in that it chews its food. Most lizards crush the food in their jaws and swallow it whole.

Mouth mash

Galapagos land iguana (*Conolophus subcristatus*)
Galapagos
Length: 4.5 feet (1.4 meters)
Land iguanas eat cacti, which provide moisture. But first, they remove the spines, either with their claws or by mashing the leaves in their mouths.

Blink

Slender glass lizard (*Ophisaurus attenuatus*)
Central, southern, and southeastern United States
Length: 3.5 feet (1.1 meters)
These legless lizards look like snakes. But unlike snakes, slender glass lizards can move their eyelids and blink. Also, unlike snakes, they have ear openings on the sides of their head. Like most lizards, glass lizards, when alarmed, can drop off their tail. The tail keeps twitching for a little while, which may tempt a predator to stop to eat it, giving the lizard a chance to escape. But this is a costly defense. The lizard's body must put a lot of energy into regrowing a tail.

Drink

Texas horned lizard (*Phrynosoma cornutum*)
Southern, central, and western United States and Mexico
Length: 7 inches (18 centimeters)
Sharp horns line the back of a horned lizard's head, and scales cover the lizard's back, sides, legs, and tail. Tiny tube-like channels between the scale edges lead to the lizard's mouth. Water that lands anywhere on the lizard's body moves into the channels and is then pulled to its mouth by the lizard opening and closing its jaws.

Cross water, tail dash

Green basilisk lizard (*Basiliscus plumifrons*)
Costa Rica, Panama, Honduras, and Nicaragua
Length: 2.5 feet (76 centimeters)
Basilisk lizards have wide, fringed toes that allow them to run across the surface of water. Holding their tails up as they dash helps them balance.

Hide

Northern chuckwalla (*Sauromalus ater*)
Southwestern United States and Mexican deserts
Length: 1.5 feet (46 centimeters)
The chuckwalla lives in rocky places. A plant eater, it hides from predators in cracks between rocks. There, it gulps in air, puffing its chest so it's wedged in tightly and can't easily be removed from a crevice.

Swim side-to-side

Marine iguana (*Amblyrhynchus cristatus*)
Galapagos
Length: 5 feet (1.5 meters)
Marine iguanas can stay underwater for up to an hour. They eat seaweed and other types of algae, which they scrape from underwater rocks with their teeth. When swimming, a marine iguana holds its arms and legs close to its body. To push itself through water, it swishes its tail side to side.

Cool your feet, two-by-two

Shovel-snouted lizard (*Meroles anchietae*)
Namib Desert
Length: 4 inches (10 centimeters)
The shovel-snouted lizard is active during the day. When the sand heats up, the lizard raises its tail. To cool off its feet, it lifts two at a time—front right with back left and front left with back right. This routine is called a "thermal dance."

Hatch

Veiled chameleon (*Chamaeleo calyptratus*)
Southwestern Saudi Arabia and Yemen
Length: 24 inches (61 centimeters)
Veiled chameleons live in trees, shrubs, and bushes. They eat mostly insects, which they catch with their sticky tongues. Chameleons' eyes move independently, so they can look in two different directions at the same time. Like most lizards, veiled chameleons hatch from leathery eggs.

Match

Satanic leaf-tailed gecko (*Uroplatus phantasticus*)
Madagascar
Length: 6 inches (15.2 centimeters)
Horns above its eyes give the satanic leaf-tailed gecko the devilish part of its name. A camouflage expert, this nocturnal lizard is nearly invisible during the day. Its tail looks like a dry, brown leaf, and its curvy body blends in with branches.

Pretend

Thorny dragon (*Moloch horridus*)
Australia
Length: 8 inches (20 centimeters)
Thorny dragons, also known as thorny devils, live in dry, sandy areas. They eat nothing but ants, which they flick up with their tongues one by one. When a thorny dragon is threatened, it can tuck its head between its front legs, showing a "false head," a knob that's on the back of its neck. This distracts attention from its real head and makes it look larger and harder to swallow.

Defend

Greater short-horned lizard (*Phrynosoma hernandesi*)
Canada, western United States, and Mexico
Length: 6 inches (15 centimeters)
This wide, flat lizard is found at high altitudes. On the back of its head are horns. When threatened, this lizard may puff up its body, hiss, and squirt blood out of its eyes!

For all the lizards and all the librarians
—APS

For Heidi, Penny, and Mochi
—SL

Acknowledgments
For keen eyes and invaluable assistance with scientific review of text and illustrations, thank you to herpetologist/author Martha L. Crump, Ph.D., of Utah State University, and Wade C. Sherbrooke, Ph.D., of Southwestern Research Station, American Museum of Natural History. Thank you also to JoAnn Early Macken and Wendy Townsend. We appreciate the International Iguana Foundation (iguanafoundation.org), the International Reptile Conservation Foundation (ircf.org), and all others who research reptiles and protect their habitats.
—April Pulley Sayre

Lizard Links

Scientists who study lizards are called herpetologists. (Herpetology is the study of reptiles and amphibians, so a herpetologist might study frogs, salamanders, snakes, turtles, and crocodilians as well.) Future herpetologists and other lizard fans can find more about these creatures in the resources below.

Books

Lizards, by Nic Bishop. New York: Scholastic, 2010.
Scholastic Reader Level 2: Lizards, by Nic Bishop. New York: Scholastic, 2014.
A Field Guide to Reptiles & Amphibians: Eastern and Central North America, by Roger Conant. Boston: Houghton Mifflin, 1998.
Sneed B. Collard III's Most Fun Book Ever About Lizards, by Sneed B. Collard III. Charlesbridge, 2012.
DK Eyewitness Books: Reptile, by Colin McCarthy. DK Children, 2012.
A Field Guide to Western Reptiles and Amphibians, by Robert C. Stebbins. Boston: Houghton Mifflin, 2003.

Websites

The Lizard Lab at Macquarie University in Sydney, Australia. whitinglab.com

Boyds Mills Press
An Imprint of Highlights
815 Church Street
Honesdale, Pennsylvania 18431
boydsmillspress.com
Printed in China

ISBN: 978-1-62979-211-8
Library of Congress Control Number: 2018940080

First edition
10 9 8 7 6 5 4 3 2 1

The text is set in Eurostile STD.
The illustrations are done in digital media.